MY NONFICTION DECODABLE READERS

Flit, Flap, Plop: A Weather Calendar

BY KIM THOMPSON

A Little Honey Book

Crabtree Publishing
crabtreebooks.com

Tips for Teachers and Caregivers

This book supports early readers as they decode words to learn facts and gain knowledge about the world.

Before reading, make sure students understand the sound-spelling correspondences shown below as well as the high-frequency words shown on the next page. Introduce the vocabulary words.

During reading, provide feedback and encouragement as students sound out decodable words by blending individual sounds.

After reading, talk about and write about the topic. Share the information on page 16 to help students learn more.

Letters and Sounds

New:

consonant blends with *l*, including *bl, cl, fl, gl,* and *pl*

Review:

all consonant sounds and standard spellings; *short a* spelled *a, short e* spelled *e, short i* spelled *i, short o* spelled *o, short u* spelled *u*

Decodable Words

add, and, at, black, blank, blasts, bliss, block, clap, class, clumps, flag, flaps, flecks, flip, flit, flock, flop, fluff, glad, glass, glints, glut, had, has, in, is, it, kids, on, plan, plants, plop, plot, plump, runs, stand, sun, up, will, wind

High-Frequency Words

New: are, brown, down, many, some, where, white

Review: a, be, days, how, live, look, of, old, out, play, the, they, time, to, water

Vocabulary Words

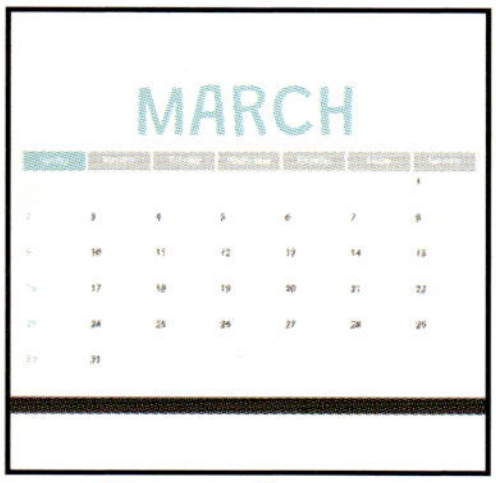

calendar

clouds

leaves

raindrops

weather

A class has a plan.

They will look out the glass.

They will look at the **weather** where they live.

They will plot it on a blank **calendar**.

calendar

On some days, flecks flit down.

White fluff clumps up.

Kids are glad to be in.

On some days,
wind blasts.

The flag flaps.

Old, brown **leaves**
flip and flop.

leaves

On some days, black **clouds** block the Sun.

Plump **raindrops** plop.

A glut of water runs.

clouds
raindrops

On some days, the Sun glints.

Plants stand up.

Kids flock out to play. It is bliss!

The plan is up.

The kids clap!

It is time to add the days.

How many days had:

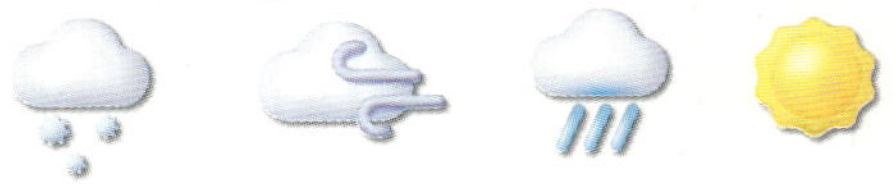 ?

Build Background Knowledge

Weather is the condition of the air over a short period of time. Weather can change from day to day, from hour to hour, and from minute to minute. Weather includes temperature, wind speed and direction, air pressure, humidity, cloudiness, and precipitation. Observing weather patterns in a place helps people know what to expect in each season of the year. Weather patterns observed over many years determine an area's climate.

MY NONFICTION DECODABLE READERS

Flit, Flap, Plop: A Weather Calendar

Written by: Kim Thompson
Designed by: Rhea Magaro
Series Development: James Earley
Educational Consultant: Marie Lemke, M.Ed.

Photographs: All images from Shutterstock

Crabtree Publishing

crabtreebooks.com 800-387-7650

Printed in China/012024/FE20231222

Published in Canada
Crabtree Publishing
616 Welland Ave.
St. Catharines, Ontario
L2M 5V6

Published in the United States
Crabtree Publishing
347 Fifth Ave
Suite 1402-145
New York, NY 10016

Library and Archives Canada Cataloguing in Publication
Available at Library and Archives Canada

Library of Congress Cataloging-in-Publication Data
Available at the Library of Congress

Hardcover: 978-1-0398-4438-4
Paperback: 978-1-0398-4519-0
Ebook (pdf): 978-1-0398-4596-1
Epub: 978-1-0398-4666-1
Read-Along: 978-1-0398-4736-1
Audio: 978-1-0398-4806-1